Evolution in Science
and Religion

THE TERRY LECTURES

Volumes Published by the Yale University Press on the Dwight Harrington Terry Foundation.

EVOLUTION IN SCIENCE
AND RELIGION

BY

ROBERT ANDREWS MILLIKAN

NEW HAVEN : YALE UNIVERSITY PRESS

LONDON : GEOFFREY CUMBERLEGE

OXFORD UNIVERSITY PRESS

THE DWIGHT HARRINGTON TERRY
FOUNDATION

LECTURES ON RELIGION IN THE LIGHT OF
SCIENCE AND PHILOSOPHY

THIS volume is based upon the fourth series of lectures delivered at Yale University on the Foundation established by the late Dwight H. Terry of Bridgeport, Connecticut, through his gift of $100,000 as an endowment fund for the delivery and subsequent publication of "Lectures on Religion in the Light of Science and Philosophy."

The deed of gift declares that "the object of this Foundation is not the promotion of scientific investigation and discovery, but rather the assimilation and interpretation of that which has been or shall be hereafter discovered, and its application to human welfare, especially by the building of the truths of science and philosophy into the structure of a broadened and purified religion. The founder believes that such a religion will greatly stimulate intelligent effort for the improvement of human conditions and the advancement of the race in strength and excellence of character. To this end it is desired that lectures or a series of lectures be given by men eminent in their respective departments, on ethics, the history of civilization and religion, biblical research, all sciences and branches of knowledge which have an important bearing on the subject, all the great laws of nature, especially of evolution . . . also such interpretations of literature and sociology as are in accord with the spirit of this Foundation, to the end that the Christian spirit may be nurtured in the fullest light of the world's knowledge and that mankind may be helped to attain its highest possible welfare and happiness upon this earth . . .

"The lectures shall be subject to no philosophical or religious test and no one who is an earnest seeker after truth shall be excluded because his views seem radical or destructive of existing beliefs. The founder realizes that the liberalism of one generation is often conservatism in the next, and that many an apostle of true liberty has suffered martyrdom at the hands of the orthodox. He therefore lays special emphasis on complete freedom of utterance, and would welcome expressions of conviction from sincere thinkers of differing standpoints even when these may run counter to the generally accepted views of the day. The founder stipulates only that the managers of the fund shall be satisfied that the lecturers are well qualified for their work and are in harmony with the cardinal principles of the Foundation, which are loyalty to the truth, lead where it will, and devotion to human welfare."

CONTENTS

EVOLUTION IN SCIENCE AND RELIGION

I.

THE EVOLUTION OF TWENTIETH CENTURY PHYSICS

IT is with very great hesitation that I have accepted the invitation to give the Terry lectures this year for I am fully conscious of the fact that I can speak with no sort of knowledge or authority in matters of either religion or philosophy. If there be any appropriateness whatever in my joining in the discussion of the relations of religion to science and philosophy it arises from two facts.

First, my life has been wholly devoted to the most fundamental of the natural sciences, physics. I have therefore had intimate contact with the spirit and with the progress of that one science and can speak with a certain amount of knowledge of its point of view. Further, I have had the good fortune to come into fairly broad contacts with the other sciences; so that I should be able to understand at least the angle of approach of scientists as such, if there be anything typical in their angle of approach, to the rest of life's problems.

Second, I have had much more intimate association as a student and teacher in three different institutions—all of which have outstanding theological departments —with the best of our religious thinkers than have most of my fellow scientists, and am therefore perhaps a little less likely to misunderstand and hence to misrepresent their point of view than some scientists might be.

In spite of these facts, my viewpoints are to be regarded as essentially individual. However incompetent any one of us may be to handle the relations of these great fields, every one of us must of necessity attempt to do so for himself if he is a reflectively moral being; for every such person must integrate his experiences into some sort of philosophy and some sort of religion. Further, as he gropes his own way—and the best of us are only gropers —he cannot possibly refuse to tell a fellow inquiring groper what he sees or thinks he sees with such light as is available to him. This is all I am trying to do in these lectures; they make no claim to authoritativeness of any sort; they represent merely my individual experience and point of view.

So far, however, as my observation goes, scientists do not differ as a class from other educated people in their attitude toward the problems of religion. This indicates, I think, not that the growth of science has not influenced religious thought, but rather that its influences are recognized in much the same way by religious leaders and by thoughtful people generally as by scientists themselves. The fact that the most outstanding scientists have frequently been men who were closely identified with religious organizations constitutes at least presumptive evidence that there is no essential conflict between the two fields; indeed, it is definite proof that there is no conflict, *as these scientists themselves* have understood and interpreted religion, for I take it that even those who are wont to make the amusing assumption that in general men who possess convictions of any sort are dishonest —that the way to become honest is to drop your principles—would hesitate to impute hypocrisy to a Maxwell or a Lord Rayleigh. I shall therefore in these lectures not focus attention upon supposed antagonisms but rather endeavor to indicate how the growth of science seems to me to have contributed to the evolution of religion in the past and what sort of influences it appears to be exerting upon its further evolution today.

In the first lecture I shall endeavor to create a background for those that follow by sketching the extraordinarily interesting and significant evolution of my own subject, physics, during the period from about 1893 when I myself first began to study it intensively up to the present. In other words, the first lecture will be pure physics. The second lecture will deal with the relations between new truth and old as it is revealed definitely in the history of physics and inferentially in other fields. This trenches somewhat at least upon the domain of philosophy. The last lecture will deal with what seems to me to have been the process of the evolution of religion under the influence of our continually expanding knowledge of the world in which we live, *i.e.,* under the influence of science.

My own period of activity in the intensive pursuit of physics happens to be almost exactly coincident with the period of development of what we may call modern physics as distinct from nineteenth century physics; so that I am in the rather unusual position of being able to relate, from my own experiences and entirely without reference to books, when and how the changes occurred, how some of the actors felt and thought and acted in the presence of each new development, and what stupendous shifts in viewpoint have been brought about. This is my excuse for making the first lecture to some extent a personal narrative.

The transition from the old to the new mode of thought in physics was probably made as dramatically in my case as in that of anyone in the world; for I was in the fortunate position of having entered the field just three years before the end of the complete dominance of nineteenth century modes of thought. In those three years I had the privilege of personally meeting and hearing lectures by the most outstanding creators of nineteenth century physics —Kelvin, Helmholtz, Boltzman, Poincaré, Rayleigh, Van't Hoff, Michelson, Ostwald, Lorentz—every one of whom I met and heard between 1892 and 1896. In one of these lectures I listened with rapt attention to the expression of a point of view which was undoubtedly held by most of them—indeed, by practically all physicists of that epoch; for it had been given expression more than once by the most distinguished men of the nineteenth century.

The speaker had reviewed, first, the establishment and definite proof of the principles of mechanics during the seventeenth and eighteenth centuries culminating in La Place's great *Mécanique Céleste;* then he had turned to the wonderfully complete verification of the wave theory of light by Young and Fresnel, between 1800 and 1830, experiments which laid secure foundations for the later structure known as the physics of the ether, one of the most beautiful products of nineteenth century thinking and experimenting; then he had traced the development in the middle of the century of the greatest and most fundamental generalization of all science, the principle of the conservation of energy; then he had spoken of the establishment in the first two decades of the second half of the century of the second law of thermodynamics, the principle of entropy or of the degradation of energy, and finally of the development by Maxwell of the electromagnetic theory and its experimental verification by Hertz in 1886, only seven years earlier than the date of the lecture. This theory abolished in all particulars except wave length the distinction between light and radiant heat and long electromagnetic waves, all these phenomena being included under the general head of the physics of the ether.

Then, summarizing this wonderfully complete, well-verified, and apparently all inclusive set of laws and principles into which it seemed that all physical phenomena must forever fit, the speaker concluded that it was probable that all the great discoveries in physics had already been made and that future progress was to be looked for, not in bringing to light qualitatively new phenomena, but rather in making more exact quantitative measurements upon old phenomena.

Just a little more than one year later, and before I had ceased pondering over the afore-mentioned lecture, I was present in Berlin on Christmas Eve, 1895, when Professor Roentgen presented to the German Physical Society his first X-ray photographs. Some of them were of the bones of his hand, others of coins and keys photographed through the opaque walls of a leather pocket-book, all clearly demonstrating that he had found some strange new rays which had the amazing property of penetrating as opaque an object as the human body and revealing on a photographic plate the skeleton of a living person.

Here was a completely new phenomenon—a qualitatively new discovery and one having nothing to do with the principles of exact measurement. *As I listened and as the world listened, we all began to see that the nineteenth century physicists*

*had taken themselves a little too seriously,
that we had not come quite as near sound-
ing the depths of the universe, even in the
matter of fundamental physical principles,
as we thought we had.*

This was the dramatic introduction,
from the standpoint of one of the very
young stage assistants in the play, to the
new period in physics. Nobody at that
time dreamed, however, what an amazing
number of new phenomena would come to
light within the next thirty years, or how
revolutionary, or, better, how incomprehensible in terms of nineteenth century
modes of thought, some of them would be.
But, at any rate, Roentgen's discovery
began to prepare the mind for the start-
ling changes that were to come. I shall
catalogue some of the most significant of
these changes under eight different heads
taking the discovery of X-rays as the first.

Second. Roentgen's discovery furnished
an instrument and a technique which made
possible the rapid development of the
electron theory of matter—one of the
grandest, because the simplest, of all phys-

ical generalizations. Although this is in a sense the very heart and soul of the new physics, I shall pass over it here with only such mention as is necessary to give it a place in the catalogue of great, new developments; because, superficially at least, the electron theory did not, at first, set itself in opposition to nineteenth century points of view. It represented the discovery of a wonderful new world, the sub-atomic world of extraordinary simplicity and orderliness, but it left the world of large scale phenomena, the old "macroscopic" world which we had known before, functioning pretty much in its nineteenth century fashion.

Third. Within a year of Roentgen's discovery, namely, in 1896, there came the discovery of radioactivity, and with that discovery, as soon as its significance began to be seen, man's view of the nature of this physical world changed overnight. Matter had theretofore been put up in a definite number—we knew not how many—of eternal, unchangeable chemical elements. In radioactivity we found two of these elements first spontaneously shooting off parts of themselves with speeds comparable with the speed of light—speeds which nobody had ever dreamed that matter in any form could under any circumstance attain—and second, by virtue of this process, transforming themselves into new elements; so that now we definitely know the life periods of a considerable number of the erstwhile "eternal" elements.

The discovery of X-rays in 1895 had revealed a whole domain of ether physics of whose existence prior to 1895 we had been completely unconscious. The discovery of radioactivity in 1896 had revealed an entirely new property of matter and quite as important a property, so far as its influence upon our conception of our world is concerned, as any which had ever been discovered. For it forced us, for the first time, to begin to think in terms of a universe which is changing, living, growing, even in its elements—a dynamic instead of a static universe. It has exerted the most profound influence not only upon physics, which gave it birth, but also upon chemistry, upon geology, upon biology, upon philosophy. Indeed, it is at this point that one of the great contributions of science to religion is now being made.

To the general public the wonder of radioactivity is now wearing off a bit merely because the phenomena have became familiar, but to the thoughtful observer the mystery is in some particulars as great as on the day of its discovery. Whence, for example, does the energy come which enables a negative electron to disregard the enormous attraction of the positive nucleus for itself and eject itself with an energy of several million volts away from that nucleus? It is just as though a huge stone instead of remaining on the earth were suddenly to decide to shoot out into space with enormous velocity against the pull of gravity. Having set up the principle of the conservation of energy as our universal guide, philosopher, and friend, we physicists of course said that either the electron which had thus ejected itself from the nucleus must have suddenly absorbed the requisite energy from some unknown ether waves which are shooting through all space, or else it must have been already endowed with an

enormous kinetic energy inside its infini-
tesimal nucleus, and some kind of entirely
unknown trigger had acted to release this
energy. The first hypothesis has already
been weighed in the balances and found
wanting, so that the second is, as we now
suppose, all that is left. Thus we saved,
after a fashion, our nineteenth century
faces, though the seeking for any kind of
mechanical model to carry the enormous
subatomic energies released in the radioactive process seems so hopeless that it has
ceased to be an interesting diversion in the
kindergarten of the physicist. *In a word,
radioactivity not only revealed for the first
time a world changing, transforming itself
continually even in its chemical elements,
but it began to show the futility of the
mechanical pictures upon which we had set
such store in the nineteenth century.*

Fourth. It may have been something of
a blow to the nineteenth century to learn
of the general transmutability of the elements, but how much more of a shock to
find that the principle of the conservation
of matter itself is definitely invalid. Beginning in 1901 the mass of an electron

was shown by direct experiment to grow measurably larger and larger as its speed is pushed closer and closer to the speed of light. But of much greater interest than that is the fact that Einstein worked out of the relativity formulae a general relation between the two quantities, energy and mass, of the form $mc^2 = E$, in which m means mass in grams, c^2 is the velocity of light squared, or the enormous number 9×10^{20}, and E is energy in ergs. This equation seems now to have the best of experimental credentials. If it is a correct one, it means that matter itself in the Newtonian sense, the quantitative measure of which is mass or inertia, has entirely disappeared as a distinct and separate entity, as an invariant property of any system. In other words, matter may be annihilated, radiant energy appearing in its place; and in view of the enormous value of the factor 9×10^{20}, a very small number of grams of matter may transform themselves into a stupendous number of ergs of energy.

It is well known with what joy the astronomers have seized upon this fact to

enable them to escape their otherwise in-
superable difficulties encountered because
the sun, for example, cannot possibly have
been pouring out heat as long as it is now
known to have been doing, if it is merely
a hot body cooling off. If, however, it has
the capacity at the enormous temperatures
existing in its interior, say 40,000,000° C.,
of transforming its very mass into radiant
energy then these particular difficulties
disappear. But what a shock it would be to
Lord Kelvin if he should hear the modern
astronomers talking about the stars radiat-
ing away their masses through the mere
act of giving off light and heat! And yet
this is now orthodox astronomy.

And, again, if they do so in accordance
with the Einstein equation then is it not
more than probable that the process is also
going on somewhere in the opposite sense
and that radiant energy is condensing
back into mass, that new worlds are thus
continually forming as old ones are dis-
appearing? These are merely the current
speculations of modern physics, based,
however, upon the now fairly definite dis-

covery that conservation of matter in its
nineteenth century sense is invalid.

Some time ago I was one of the speakers
at a forum, and in the course of my ad-
dress I used the word "spirit" a number of
times. When questions were afterward
called for, a man arose in the rear of the
room and with a somewhat hostile air
asked if the speaker would define what he
meant by the word "spirit." I replied that
if the interrogator would be good enough
to define for me the word "matter" I
would attempt to define for him the word
"spirit." The attempt was not called for.
And, in fact, in view of the growth of
twentieth century physics and the changes
in our conception of matter that it has
brought, it is today quite as difficult to
find a satisfactory definition of "matter"
as of "spirit."

Fifth. But what do we now know about
the nature of this phenomenon which we
have called radiant energy, with the aid of
which the masses of the stars are being
dissipated into space? In a word, where is
now the nineteenth century physics of the
ether?

The physics of the ether meant in 1890 the physics of electromagnetic waves, and it means precisely that *now*. Electromagnetic waves are sharply and definitely recognizable by certain observed properties. Thus, in the first place, electro-magnetic waves travel through space with an exactly measurable speed, namely, the speed of light, *i.e.,* 186,000 miles per second. Second, they all exhibit a definite, measurable periodicity, or frequency, which, divided into the velocity of light, gives the wave-length. Third, they all exhibit another measurable property described by the words "state of polarization," the precise definition of which need not here be given. Note that these properties are completely independent of all theories as to the nature of electromagnetic waves.

We can produce and study electromagnetic waves of an infinite variety of frequencies ranging from very long wireless waves, kilometers long, up through heat-waves to light-waves of wave length of the order of $\frac{5}{10000}$ mm., up still farther through ultra-violet rays to X-rays of frequencies 10,000 times that of ordinary light, and up again through gamma rays to the cosmic rays of frequencies again 1,000 times those of X-rays. All this to show that the physics of the ether is not a vague set of ideas, but that it deals with sharply measurable experimental facts, the validity of which is unquestioned, and which are completely independent of all speculations and theories.

Now if in 1890 any physicist had been asked to describe the mode of interaction between ether waves and, say, electrons in atoms, he would presumably have answered with great definiteness and assurance about as follows: "It is essentially the same as the mode of interaction between a tuning fork, or a piano string, and the air waves produced by its vibration. The fork sends out into the surrounding air a series of waves the period of which, of course, synchronizes with the period of the vibrating prong. If just such a series of waves should fall upon the fork when at rest it would pick up these waves and be set into vibration by them. But this would be true when and *only* when the frequency of the impressed or oncoming waves coincides with the natural frequency of the tuning fork. Precisely similarly the electrical charges—now called electrons—which are in the atoms of matter when these atoms are, for example, in the suns and stars, are set into all sorts of rapid vibrations, and these vibrating electrical bodies impress their individual, or, better, their integrated, wave form upon the ether just exactly as the instruments of the orchestra impress their integrated wave form upon the air which transmits it to your ear."

Now up to 1900 all the phenomena of ether waves had seemed to fit accurately and beautifully into this sort of wave theory. Its successes were almost countless. It explained beautiful and intricate phenomena like the colors of soap bubbles and interference patterns of even the most complicated sort. Up to 1900, I say, the theory had never failed. But during the first fifteen years of this century there was discovered a group of new phenomena which baffled explanation in terms of nineteenth century ether physics. These phenomena are as follows: When ether waves of sufficiently high frequency are allowed to fall upon the atoms of matter they are found to jerk the electrons from these atoms, and in so doing to communicate to them a kinetic energy which is independent of the intensity of the incident waves, but is accurately proportional to their frequency; *i.e.,* kinetic energy imparted = $E = h\nu - p$ where h is a universal constant, ν the frequency of the incident waves, and p the work necessary to detach the electron from the metal. This is a phenomenon that has been checked in all sorts of ways and has always and everywhere been verified, but it is one which is to the present day completely inexplicable in terms of the nineteenth century wave theory. It obviously fits better some sort of corpuscular theory than a wave theory. It is a new phenomenon of stupendous importance for the understanding of the foundations of the physical world in which we live.

Sixth. But now came, about 1913, the discovery of the effect inverse to the last. Not only was the energy communicated to an electron by an ether wave, which was absorbed by that electron, proportional to the frequency of the wave, but when atoms of substances like glowing hydrogen, for example, emit ether waves, the frequency of the emitted light can be found by considering the electron in the act of emission to have fallen from one energy level E_1 to a second E_2 and to have emitted a frequency proportional to the change in energy, *i.e.*, to $E_1 - E_2$, the factor of proportionality being the same universal constant h; so that the equation $E_1 - E_2 = h\nu$ gives a reciprocal relation between the electronic energy and ether wave energy. This experimental discovery, first that the frequency of an ether wave may be taken as a measure of its energy available for absorption by electrons, and second that the energy $h\nu$ can interplay in this way between ether waves and electrons in atoms, is so irreconcilable with the wave theory of the nineteenth century that Einstein has suggested abandoning the wave theory of light altogether and returning to a modified corpuscular theory of the transmission of radiant energy through space.

Also at the hands of A. H. Compton this new light-dart theory has recently had new and striking success, but nobody has as yet been able to show how such a light-dart theory can account for the scores of interference phenomena so beautifully explained by the wave theory. Such is the *impasse* confronting physics today in its endeavor to obtain a picture of the mechanism of the transmission of radiant energy through space. We have discovered a whole group of new phenomena of radiation to which the old laws do not apply; yet we must retain the old laws for the interpretation of the old phenomena.

Seventh. Not only are we at present completely unable to form any consistent picture of the mechanism of the transmission of radiant energy, but new experiments have recently come to light which show conclusively that the frequency of an ether wave is not produced by, and does not correspond to, a synchronously vibrating electronic tuning fork within the atom at all. *We can at present make no mechanical picture whatever of the act by which an ether wave is born and started out on its journey through space.* Two electrons within the same atom have been definitely found in some instances to jump simultaneously each to a new position, and the sum of the energies of these two changes are somehow integrated by the atom into a single monochromatic ether wave of a frequency corresponding to the sum of the two electron jumps. In a word, of the process by which an ether wave is born we know only this much, that every atomic shudder (change in energy) of whatever sort seems to become integrated into a monochromatic ether wave, the frequency of which is computable from $h\nu = E_1 - E_2$, E_1 being the atomic energy before the shudder and E_2 that after it, so far have we got from the simple mechanical picture of a little electrical vibrating tuning fork sending off waves into the ether synchronously with its own vibration! *Both the mode of birth of an ether wave by an atom, and its mode of transmission from star to star after birth are still almost complete mysteries.*

Eighth. I shall mention but one more of the large category of discoveries constituting twentieth century physics. It is perhaps the most striking and revolutionary of them all, the discovery that the very foundations of mechanics when looked at microscopically are unsound, that apparently all periodic motions are resolvable into circular and linear coördinates which cannot progress continuously as demanded by the Newtonian laws, but are built up out of definite unitary elements; specifically, that a body rotating in a circle can possess only such angular momenta as are exact multiples of a universal unit of angular momentum, *viz.*, $\frac{h}{2\pi}$. This unitary fine structure in motion, like the unitary fine structure in electricity, we had never discovered, or even dreamed of, until this century, because we had never experimented upon small enough angular momenta on the one hand, electrostatic charges on the other, to see that each had in fact a granular structure. When one is weighing sand by the ton it has for him no granular character. It is only when he begins to weigh quantities of the size of individual grains that he sees it to be granular. That periodic motion itself has such a granular nature is one of the most amazing experimental discoveries of our century. *We can still look with a sense of wonder and mystery and reverence upon the fundamental elements of the physical world as they have been partially revealed to us in this century. The childish mechanical conceptions of the nineteenth century are now grotesquely inadequate.*

We have at present no one consistent scheme of interpretation of physical phenomena, and we have become wise enough to see, and to admit, that we have none. We use the wave theory, for example, where it works; we use the quantum theory where it works; and we try to bridge the gap between the two apparently contradictory theories, in purely formal fashion, by what we call the correspondence principle. *It is true that we are slowly learning more of the rules in nature's game,* so that our progress is not made by hit or miss experimenting, nor by random theorizing, but by following a more or less systematic, if not always a strictly logical, procedure; but the day has gone by when any physicist thinks that he understands the foundations of the physical universe as we thought we understood them in the nineteenth century. The foregoing discoveries of our generation have taught us a wholesome lesson of humility, wonder, and joy in the face of an as yet incomprehensible physical universe. *We have learned not to take ourselves as seriously as the nineteenth century physicists took themselves. We have learned to work with new satisfaction, new hope, and new enthusiasm because there is still so much that we do not understand, and because, instead of having it all pigeon-holed as they thought they had, we have found in our lifetimes more new relations in physics than had come to light in all preceding ages put together, and because the stream of discovery as yet shows no signs of abatement.*

NEW TRUTH AND OLD

II.

NEW TRUTH AND OLD

THE more intimately one gets into
touch with any civilization of by-gone
days and the more carefully he studies the
thought, the feelings, the action of men
and women who lived in Pompeii, in
ancient Athens, in Thebes, or in Babylon,
the more is he struck by the similarity between the way people seem to have lived
and talked and thought two or three
thousand years ago and the way they live
and talk and think now. The jokes and the
horseplay of Plautus are surprisingly like
the humor of *Puck,* or *Life,* or *Punch,*
or *Fliegende Blätter.* One can imagine
Cicero, the wisest of the Romans, sitting
with his friends in a patio in Pompeii and
saying nearly the same things we say as
we sit in patios in California or Florida.
The beauty of women, the strength of
men, the flavor of strawberries, the delec-
tability of oysters, the horrors of war, the
glory of the landscape, the aroma of

flowers, the love of friends, courtship,
marriage, and divorce, the racetrack, the
wrestling match, the boxing bout—all
these played almost exactly the same rôle
in the lives of the people of Rome as they
play now in the lives of the people of New
Haven or New York. And it is around
these things, too, that about 90 per cent of
the interests of the average man revolve.

Even in what are called the higher
things of life can we truthfully be said to
have made, or to be making, any real
progress? That question has often been
raised and sometimes answered negatively
by literary men of reputation, and sometimes even by philosophers. In Tut-ankh-
amen's tomb are found evidences of artis-
tic development three thousand years old
quite the equal of our own in similar lines.
Greek sculpture and Greek architecture
we can but imitate today. In intellectual
power we do not surpass, even if we equal,
the Athenians or the Alexandrians. In
devotion to moral and spiritual ideals,
where can the twentieth century show
anything finer than the sublime story of
the death of Socrates? And was it not

2000 years ago in Galilee that one lived of whom the whole thinking world still says "never man spake as this man"? Was the old cynic of Ecclesiastes right when he said, "Is there any thing whereof it may be said, See, this is new? It hath been already of old time, which was before us."

It is that question which I wish first to try to answer, "Is there any thing whereof it may be said, See, this is new?" and I propose to answer it by describing briefly the birth of two ideas, one of which is so young that I myself have actually lived through the whole period of its birth and development, and yet which has already become so significant that it is not too much to say that it marks a new epoch in civilization. The other idea is indeed already grown to maturity, though it has certainly not "been of old time." It is about 300 years old, but in that 300 years it has probably exerted a larger influence upon the destinies of the human race than has any other idea which ever entered the human mind. The story of its birth is an intensely fascinating one.

Imagine a world in which civilizations,

and some of them of an extraordinarily high order, have been in existence for five or six thousand years at least, possibly much longer—a world in which men have thought about all human relationships much as we think now—have pondered just as penetratingly as anyone has ever pondered over the meaning and nature of existence, over the questions, "What is truth?" "What is reality?"

Imagine a world in which billions of people have already lived and died, people of just as high intelligence as those now living, people who have exercised their intelligence as most of us do now, mainly upon earning a living, but people among whom there have yet developed in many places, notably in Greece, outstanding thinkers who have set themselves the colossal task of trying to understand "The Whole." Occasionally, too, one has arisen who was willing, now and then, to turn aside, as Aristotle did, from the contemplation of the problem of the ultimate reality to set himself a simpler question, and to ask whether we understand thoroughly just some little part of how nature works, for example, a question like this: What is the *natural state* of bodies on the earth with respect to motion?

Imagine a world which had always answered that particular little question whenever and wherever it had been raised, as follows: The natural state of bodies with respect to motion is of course a state of rest. Every body living or dead comes to rest when it ceases to exert itself or to have effort spent upon it, and the force or effort that must be exerted upon a body to get it out of that state is, of course, proportional to the amount by which it is made to depart from its natural state, that is, it is proportional to the velocity imparted to it. So long as we are thinking about kicking a stone, or throwing a discus, or hurling a javelin, or pushing a cart, or drawing a chariot, or about almost any of the common phenomena involved in the production of motion, the foregoing answer is a natural, and an almost inevitable one. Indeed, it is strictly correct for all sufficiently slow motions through resisting media. The foregoing answer had been given for at least fifteen hundred years, and everybody who had thought about it at all had been altogether content with it. You and I, had we lived then, would doubtless have given it our approval.

But about 1560 a man was born who began to form the habit of concentrating his attention upon less colossal and more detailed problems than had occupied the thought of most of his predecessors. He first sets himself this old problem, as to what is the natural state of bodies with respect to motion, and he begins to think all around it to see if all the motions of which he can think are fitted by the old formula, and he finds one that does not seem to fit. He reflects that one body that is twice as heavy as another does not seem to fall from the table to the floor in half the time. To obtain more precise, quantitative evidence he goes to the top of the Leaning Tower of Pisa and drops his two bodies of like dimensions, one of wood and one of metal. They strike the earth practically simultaneously, that is, they acquire the same velocities, though one is pulled toward the earth with a force many times that acting upon the other; and the rule that men had accepted for thousands of years on qualitative or hearsay evidence is gone forever.

But this is not the birth, nor even the conception of the idea of which I am speaking. Thales of Miletus, 600 years B.C., had done experimenting on amber; Archimedes of Syracuse by a flash of intuition which we moderns call a "hunch" had discovered, about 225 B.C., a great hydrostatic law while he was in his bathtub; Aristotle of Athens had lauded the experimental method. But now, Galileo, unlike any of his predecessors so far as I can discover, began to apply the scientific method with an intensity and in a way that was altogether new. He set about devoting his life to the task of devising a long series of experiments for obtaining quantitative evidence which might enable him to replace the old wrong idea by the correct one—experiments involving the introduction of new and accurate methods of measuring both times and velocities. His mode of attacking his problem was entirely new, and as a result of his long series of careful experiments he proved conclusively that the natural state of a body is *not* one of rest, but that it is rather a state of rest *or of motion in a straight line;* and that force or effort is proportional, not to motion, but to the rate of change of motion. In other words, he got the idea that is expressed in the famous equation of motion of Newton's, $f = m\,a$.

The idea that was then born, however, was not so much that embodied in the result $f = m\,a$, as in the method. True, the equation itself is of well-nigh infinite importance. Not a single dynamical machine in existence today can be designed without its aid, not a steam engine, not an automobile, not a dynamo, not a motor, not an aeroplane—not a machine or device of any sort for the transformation of work or for the utilization of power. Subtract merely the result $f = m\,a$ from modern civilization and that civilization collapses like a house of cards, and mankind reverts at once to the mode of life existing in ancient times, a life well equipped with statics, such as are involved in building processes, but wholly wanting in dynamics. But it is not that fact that gives the idea its highest credentials.

It is true, too, that this equation changed man's whole conception of the universe, though it took at least two centuries to accomplish that result. For this was the indispensable idea that made possible, first, the discovery of the law of gravitation, and, second, the ultimate triumph of celestial mechanics—a triumph which robbed the planets of their age-long dominance over the lives of man, of their proud position as arbiters of his destiny. Other ideas contributed to that end, but this was the corner stone. It was two centuries before the new idea found any practical application to the affairs of earth, *i.e.,* before the laws of celestial mechanics began to be applied in the development of terrestrial mechanics, but during all that period it reshaped philosophy and it reshaped religion, *for through it mankind began to know a God not of caprice and whim, such as were all the gods of the ancient world, but a God who works through law.* It also changed conceptions of duty, the second element in religion, because it began to reveal a nature of orderliness, and a nature capable of being known; a nature, too, whose functioning might be predicted, a nature which could be relied upon; a nature, also, of possibly unlimited forces, capable of being discovered, and then of being harnessed for the benefit of mankind.

To get the meaning and the effect of this new idea, compare the monasteries of the middle ages filled with serious souls, who, finding a bad world, saw nothing to do about it except to escape from it and cultivate their souls; compare all this with the incessant activity in the service of humanity of a Kelvin or a Pasteur. These new conceptions, first, of the creation, or of the creator by whatever name you may wish to call him, and, second, of human responsibility, of man's place in the scheme of things—the two elements of all religion—came into human thinking and began seriously to influence human conduct about Galileo's time. There are glimpses of them in the adumbrations of the Greeks, but they first began to spread and to affect in a large way human life about Galileo's

time, and as a result of his work and that
of his contemporaries and followers. But
it is not even in these services that the idea
of which I speak finds its chief credentials.

It was rather the *method* used by Galileo, and followed by Newton and Franklin and Faraday and Maxwell and Pasteur and Darwin, and a host of others
who caught its significance; it was this
that constituted the great new idea—an
idea which has finally brought into the ken
of mankind the conception of an evolving,
developing, progressing world. Through
it we have learned to read the story of the
geological evolution of the earth, of the
evolution of *life* on the earth, of human
history and civilization, of the stars themselves, and even of the elements of which
the stars are made.

Is it not the most sublime, the most
stimulating conception that has ever entered human thought, this conception of
progress, this new idea absolutely unknown in ancient times, a progress of
which we are a part, and in which we are
ourselves consciously playing a rôle of

supreme importance?—a progress which, whatever doubts Dean Inge and Professor Bury may give voice to, the man on the streets now partially understands and certainly believes in; for has he not, in his own lifetime, seen the most capricious, the most terrible of the phenomena of nature, the thunderbolts of Jove, become the willing slaves of all mankind, so that six million people in America alone are today supported directly or indirectly by the electrical industry? If we define progress as the increasing control over environment, then it is something which the average man thinks has been definitely revealed to him through the study of the history of organic life on the earth, and which has certainly been revealed to him through the study of human history, particularly nineteenth century history. You may call it an illusion if you wish. We may admit that there is no proof that it will go on forever. But if the average man finds that it has already been going on for millions upon millions of years he is not much concerned about a mere dialectic discussion of the term forever. A few more millions of years is enough eternity for his purposes.

Certainly through the method of Galileo, and the success that its pursuit has already brought, mankind has just recently begun to glimpse limitless possibilities ahead of it in the understanding of nature, and in the turning of her hidden forces and potentialities to the enrichment of life. Nobody knows to what limits we shall be able to go with the aid of this method, but if the past three hundred years is an index of what the next three hundred years may be, then *the supreme question for all mankind is how it can best stimulate and accelerate the application of the scientific method to all departments of human life.*

What, now, is the second idea, the one that is just being born? Galilean and Newtonian mechanics, with all their incomparable consequences for the material, the intellectual, and the spiritual life of the race, developed with ever-increasing effectiveness through three centuries until toward the end of the nineteenth century we began to look upon them, as I pointed out in the last lecture, as final verities, practically incapable of extension or expansion. All the great foundation principles underlying the physical world were said to have been discovered. New applications of course would be found, but matter, defined by the property of inertia—the m appearing in the equation $f = m\,a$— would everywhere be found to be conserved. Further it had been found to be put up in seventy-odd fundamental packages, called the chemical elements, all of which were unchangeable and eternal.

And outside the physics of matter there was the physics of the ether, completely described and governed by the equations of Maxwell, also thought by most physicists to be finalities, the four watch dogs that kept the whole works in order being the principles of (1) conservation of matter, (2) conservation of momentum (Newton's second law), (3) conservation of energy, and (4) Maxwell's equations. This was the situation at the time when I myself and others of my age began the intensive study of physics.

And then, through the continuous use

of the method of Galileo, during the last five years of the nineteenth century the new idea began to be born, an idea which has created what we call modern physics. And this in its turn has already made this particular generation in which you and I are living, this last thirty years, from the standpoint of the importance of the discoveries that have been made and of the fundamental character of the changes in our conceptions that have taken place, more significant than any of the generations of the past, or than all of them put together, with the possible exception of the generation of which I have just been speaking.

With the precise character of these changes and how they have been brought about I shall not now deal, for I have treated these in the preceding lecture, but I wish here to christen this new idea. I shall call it "the idea of the electrical constitution of matter," and I wish here, too, to restate one or two of the most important aspects of the advance that is just now being made in so far as one can yet gauge it.

In the first place, the dogma of the immutable elements is gone, forever gone. It went with the discovery of radioactivity. In that discovery even the physical world changed in our thinking overnight in its fundamental elements from a fixed, changeless, static, dead thing to a changing, evolving, dynamic, living organism.

In the second place, the two fundamental principles, conservation of mass and conservation of energy, are now gone *as distinct and separable verities,* as each in itself a universally applicable proposition; and Maxwell's electromagnetic equations have suffered a like fate. We have actually found instances in which mass seems to be transforming itself into energy, as well as energy into mass: in other words, we have seen the conceptions of the conservation of energy and the conservation of mass become completely scrambled. Also we have found in quantum mechanics situations in which both the principle of conservation of momentum and the validity of the Maxwell equations have had to be given up.

But perhaps someone says, What of it?

Of what practical use were these abstract conceptions anyway? Then let me give just one more illustration of the results which have already sprung from the birth of the new idea. For about fifteen years, instead of two centuries, as in the case of Galileo's discovery, the knowledge of the electrical constitution of matter was worked out by the methods of pure science in the laboratories mainly of college professors, who were intent only upon increasing our understanding of electronic phenomena, confident that their job was worth while, but not interrupting their quest to try to take out patents or to seek industrial applications. In this way a large body of knowledge about the characteristics of electronic discharges in exhausted tubes was developed. About 1910 the demand was keenly felt for the development of a telephone relay or repeater in order that long distance wire telephony might not be too costly for commercial realization, and the inertialess characteristics of these electronic discharges made them peculiarly fitted for this purpose. It was 1914 before the first electron tube repeaters or amplifiers were installed on commercial lines. The enormous strides made within the past dozen years in the field of communications, the whole art of broadcasting, the whole increase of the efficiency of telephone lines because of the use on them of carrier wave frequencies, nine-tenths of the immense development in the whole field of speech transmission, a series of developments which make a business in the United States alone running into a billion dollars annually—all this is due to the utilization in the vacuum tube repeater or amplifier of the knowledge of electronic phenomena first worked out in the pure science laboratory. And the mere dollar values of these developments are as nothing compared with the values some of them already have, springing from the amelioration of the conditions of country life, and the educational possibilities for the whole communty which are here foreshadowed.

Is it any wonder the physicist feels that the birth of the second idea is likely to be of quite as stupendous moment for the destinies of the race as the birth of the first has already proved itself to be? Is it surprising that he has such confidence in the altogether limitless possibilities ahead for the enrichment of life and for the future development of a better civilization, provided only we can divert a small part, a very small part, of the energies and resources of mankind from gum chewing and movies, and joy riding and cosmetics and indulgences of all sorts to the furtherance of the immense task of ferreting out nature's secrets and applying them to human needs?

Thus far I have merely been giving my answer to the question, "Is there any thing new under the sun?" In so doing I have had merely to tell the story of the beginnings of physics in Galileo's time, of its subsequent development and application to human life, and then of the changes in physics going on in our own generation. That story seems on its face to justify the philosophy of revolution and of protest, for has not twentieth-century physics shown that nineteenth-century theories were merely "a pack of lies," rubbish fit only to be dumped out and

forgotten? Have we not just said that practically all the props which held up the nineteenth century structure of physics have fallen? That is precisely the way a devotee of the philosophy of protest would state it. But in so doing he would be either under a complete misconception, or else guilty of a complete distortion of the facts. For the exact and obvious truth is that no discovery of the twentieth century has thus far subtracted, nor can it ever subtract, one whit from the great body of *experimental facts* brought to light in the nineteenth century. *These facts, some of them of incalculable importance, too, are henceforth the permanent heritage of the race.* In them eternal Truth has been discovered, truth that will forever guide the race in its effort to live in better accord with, better understanding of, better control over, nature. In other words, *experimental* science never has and never can take a backward step. It moves only forward in ever-expanding circles.

Secondly, how about the framework of theory which had held these facts together, given them unity, assisting the memory in retaining them and in giving them rela-

tions? Is this all a lie? Emphatically no.
The seventy-odd elements of chemistry,
now become ninety-two, are still and
always will be the ultimate units of chemical combination, which was the only field
in which their ultimateness had been
tested before 1895. In the study of chemistry and in its practical applications, too, we
use the elements now precisely as we used
them before the discovery of the electron.
The only change is that the physicist has
discovered a new field, entered by new
physical processes—X-ray processes,
radioactive processes, cathode ray processes, in which the ultimateness of the
elements disappears. *This means that we
have merely had to supplement, extend,
build over a bit our old theories, not
abandon them.* It is amazing how seldom
a well-worked-out physical theory is
abandoned. The principle or theory of the
conservation of mass, instead of being fit
only for the scrap heap, still holds in all
the experimental fields to which we had
access when it was set up. We have found
it failing only when quite recently we
began to be able to investigate bodies
moving close to the speed of light. What

we have found is merely that the principle is not of as *universal* validity as we thought it was. For the practical purposes of our present work-a-day world it is completely true.

Again, the principle of the conservation of energy is still the corner stone of all engineering and of practically all physics, and that, too, in just the way in which we used it before it became mixed up with mass. It is still eternal truth for the fields which called for its formulation. The interconvertibility of mass and energy has indeed significance for us at temperatures existing inside the stars, but these are temperatures to which even in our thinking we have never until very recently had any access. Also it has very recently (1933–34) found large significance in interpreting the transmutations produced by the bombardment of certain elements with neutrons, alpha particles, protons, deutons and other sorts of atomic nuclei; but these transmutations, while they add new knowledge to our store, render not a bit of the old knowledge invalid.

Finally, the principle of the conservation of momentum and the laws of electro-

dynamics are still true in all large scale
phenomena, such as alone we dealt with up
to 1900. The whole of our industrial application of electricity is today governed by
them. It is only by learning to peer inside
the atom that we have discovered a new
region in which there are exceptions to
these laws. Even the revolutionary discoveries in the domain of quanta have not
yet enabled us to give up the wave theory.
We use it side by side with the quantum
theory in spite of the apparent irrecon-
cilability of the two, simply fusing them
together for the purposes of mathematical
description and prediction into what we
call The New Wave Mechanics —a beautiful illustration of the fact that *science
has little to say about ultimate causes.* Its
concern is the observation of phenomena,
and the fitting of them together into as
comprehensive a theory, or theories, as it
can find, primarily for the sake of predict-
ing new facts, to be in their turn subjected
to the test of new experiments. It built up
in precisely this way a wonderfully complete body of definite knowledge about
electricity in conductors before we knew
anything about the electron or about the

relations of electricity and matter which
have recently come to light. Similarly, it
built up an immense field of definite
knowledge about ether physics before
matter and radiant energy had been
linked together at all.

In a word, nineteenth century physics
was in 1900, and it is today, eternal truth
so far as applications to the domain of
knowledge to which we had access in 1900
is concerned, and what is true of physics
is probably to a very large extent true of
the growth of all knowledge. The civilizations of the past have all of them discovered truth. Some of them, perhaps
most of them, have gone as far as they
could *with the observational data with
which they had to work.*

Why is it that we have never surpassed
the sepulchral decorative art of the Egyp-
tians, nor their sepulchral architecture
either? Is it not because they, too, at least
in some of the fields in which they worked,
discovered eternal truth? Why is it that
in the plastic arts, in aesthetics, in certain
forms of the drama, in the exercise of pure
reason, we can only *imitate* the triumphs
of the Greeks? Is it not because the Greeks

discovered in these fields eternal truth? Why is it that all the world is still willing to say of Jesus, "Never man spake like this man"? Is it not because he literally spake two thousand years ago the words of everlasting life—the words of rich, full, abundant, satisfying, unselfish living for all times and all places?

The first inference, then, that I wish to draw from this survey of the growth of modern physics is that there is continually appearing "something new under the sun," and there is no reason at all to suppose that it is confined to the field which has given birth to the two particular ideas concerning which I have spoken—*that there is a truth in the past which is not, and cannot be ignored or brushed thoughtlessly aside by men of insight and understanding,* that much of the knowledge of the past is still eternal truth, that just as Einstein embraces the whole of Newton, so presumably the truth of the present is merely a supplement to, an extension of, the truth of the past. It takes on a new aspect, a richer, completer significance with every advance in knowledge, but only the undiscerning and the thoughtless fail

to see the truth that was clothed in the old
dress—fail to see, in a word, that this
whole process of which we are a part is a
slow continuing *growth*. If it fails to ap-
pear as such at times it is only because we
have not a wide enough perspective, because our candidate, for example, has been
defeated at the polls, and we accordingly
think for the moment that the whole march
of progress has been reversed. At such
times we need to reflect upon such a bit
of doggerel as appeared some years ago
in the *Outlook* just after its candidate and
its policies had suffered disastrous over-
throw.

> My grandad notes the world's worn cogs
> And says we're going to the dogs.
> His grandad in his house of logs
> Thought things were going to the dogs.
> His dad, among the Flemish bogs,
> Swore things were going to the dogs.
> The cave man in his queer skin togs
> Knew things were going to the dogs.
> Yet this is what I'd like to state,
> Those dogs have had an awful wait.

The undiscerning and the thoughtless
are divided into two great groups, the
one the conventional crowd which simply

passes on the past wthout change, the
other the red mob, the devotees of the next
easiest and cheapest philosophy, "the
philosophy of knock." The unthinking
join each of these two groups in crowds.
But the man of education and intelligence in general joins neither. *Indeed, is
not the main purpose of education to enable one to know the truth of the present,
and to understand the truth of the past;*
in a word, to enable one to estimate cor-
rectly his own place and that of his contemporaries in the ever-expanding ocean
of knowledge, for only with such understanding can he shake off the inhibitions
of the conventional, free himself from the
emotional futility of the radical, and put
forth constructive effort for the real
betterment of the world?

One or two illustrations of effort that
is not constructive will be illuminating.
After the discovery of the law of gravitation all attempts to make new physical or
engineering developments, save such as
are consistent with and limited by this law,
became, of course, ridiculous, since they
ignore fundamental and established truth.
Precisely similarly with respect to pro-

posed violations of the principle of conservation of energy and the laws of elec-
trodynamics as applied to large scale phenomena. But this ridiculousness does not
prevent inventors without background
from continually putting forward perpetual motion machines, nor does it prevent ignorant or unscrupulous persons
from advertising Abrams electronic reactions, magnetic belts, and the like.

Also such persons undoubtedly have
their exact counterparts in the fields of art,
finance, education, and all other departments of human activity, persons who are
ignorant of fundamental laws that *have
been discovered,* who are hypnotized by
anything which is new because it is *new,*
and who are not interested in first finding
what has been found to be *true;* for there
are presumably fundamental laws in art
as well as in physics, in accordance with
which all real progress must be made. One
of the foremost painters of the United
States told me recently that he considered
a very large fraction of what is called
modern art to be in the precise category of
perpetual motion machines and Abrams
electronic reactions—a violation of the

fundamental laws of real art, and hence
doomed to disappear like all other untrue
things. And how many cubists we have in
economics, in education, in government, in
religion, everybody knows,—persons who
are unwilling to take the time and to make
the effort required to find what the known
facts are before they become the cham-
pions of unsupported *opinions*—people
who take sides first and look up facts
afterward when the tendency to distort the
facts to conform to the opinions has become well nigh irresistible.

The second inference that I wish to
draw from my review of the growth of
modern physics is that there may be some
danger that we may not even yet have
learned to avoid the blunder made by the
physics of the nineteenth century. This
blunder consisted in generalizing farther
than the observed facts warranted, in
assuming that because no exceptions had
been found to the validity of the principles
of the conservation of mass, of momentum,
etc., therefore no fields would ever be
opened in which these laws failed; in a
word, *the assumption that our feeble,
finite minds understand completely the*

basis of the physical universe. This sort of
blunder has been made over and over and
over again throughout all periods of the
world's history and in all domains of
thought. It is the essence of dogmatism—
assertiveness without knowledge. This is
supposed to be the especial prerogative of
religion, and there have been many religious dogmatists, but not a few of them,
alas, among scientists. Everyone will
recognize Mr. Bryan, for example, as a
pure dogmatist, but not every scientist
will realize that Ernst Haeckel was an
even purer one. If there is anything that
is calculated to impart an attitude of humility and of reverence in the face of
nature, to keep one receptive of new truth
and conscious of the limitations of our
finite understanding, it is a bit of familiarity with the growth of modern physics.
It is quite as effective as the "tropic
forests" which put Charles Darwin into
such an attitude of reverence when he
wrote, "No man can stand in the tropic
forests without feeling that they are
temples filled with the various productions
of the God of nature, and that there is
more in man than the breath of his body."

THE EVOLUTION OF RELIGION

III.

THE EVOLUTION OF RELIGION

GILBERT MURRAY, famous scholar, humanitarian, idealist, one of the sweetest-tempered of men, when asked last summer what he thought about the Scopes trial, wrote: "The most serious set-back to civilization in all history—this is my considered judgment of the Scopes trial."

If he will let me I should like to modify the statement just a bit so as to make it correspond more exactly to my own convictions, and I think that he will not object to the change. I suspect that the idea to which he meant to give expression was that if the present effort, of which the Scopes trial is an illustration, to suppress freedom of thought and to return to the spirit and the method of the Inquisition, barring only the element of physical torture, were to be successful in the United States, it would be the worst set-back to civilization in all history. And in that statement I should quite agree with him.

Nevertheless, I am not very greatly disturbed about the present situation, because I do not see that these efforts are being successful, or that there is much reason to fear that they will be. Indeed, I am inclined to go farther and to say that the net result of the Scopes trial and of all the newspaper discussion that has gone with it has been good rather than evil. For it has set tens of thousands, hundreds of thousands, perhaps millions of people, who have not been in the habit of doing so before, to reflecting for themselves upon the basis of their own religious conceptions. In this respect the Scopes trial has probably been one of the big educative forces of the present decade, and I do not see anything but good that is coming out of it.

The great majority of us probably live and act most of the time, and all of us no small part of the time, purely conventionally. We do the things which the people around us do. We talk piously about law enforcement and the Constitution, as members of the Better-America Federation, and forget all about law enforcement and the Constitution when it becomes customary in our social group to dodge taxes, violate speed laws, and support bootleggers. And yet the basis of all character and the *sine qua non* of all progress is obviously what Tufts and Dewey call "reflective morality," as distinguished from conventional morality, if this latter can be called morality at all. And if there is anything that this world needs, it is the spread of a little reflective morality among its people, and I am inclined to think that all this discussion aroused by the Scopes trial has tended to develop it. If I can assist ever so little by presenting some of my own reflections upon the place of evolution in religion, I shall consider myself amply justified for having the temerity to speak at all in a field in which I can at best speak with no sort of authority.

I shall state my conclusion at the outset when I say that religion itself is one of the most striking possible examples of evolution. In so saying I am uttering nothing that is in any way heretical, nothing that is not said in every theological seminary of importance in every denomination in the United States, nothing that is not said in every group of people who do any reflecting at all, or who have any sort of familiarity with history and its interpretation. For nothing stands out more clearly, even in Bible history, than the fact that religion, as we find it in the world today, has evolved up to its present state from the crudest sort of beginnings, and I propose to run rapidly over four stages in that evolution.

We do not have to be anthropologists, or to have made special studies of primitive man, to see how crude a thing his religion was. We have many primitive men living today, just as we have all sorts of survivals and vestiges of former very prevalent types, and we scarcely need more than a superficial familiarity with the way the native of the Congo pounds his tom-tom to scare away the god that he fears, or with the incantations, the totem poles, and the medicine-man practices of our American Indians to justify some such picture of the beginnings of religion as the following.

Primitive man, just beginning to come

into consciousness of himself, to act not altogether instinctively as the lower animals for the most part do, but with a little bit of reflection, finds himself on the one hand surrounded by human enemies, who kill and enslave him, to whom he is obliged, if they are more powerful than himself, to surrender the best that he has, —his sons and his daughters sometimes— to make possible a cannibal feast. On the other hand, he finds himself surrounded at the same time by the forces of nature which seem to him as capricious as his human enemies. It is a nature which sometimes smiles upon him and sometimes is very angry, which strikes him down with lightning, wastes him with disease, lets him die of hunger·

Under these conditions what does he do? Probably the only thing that a man in his stage of development can do; he personifies nature. He sees a spirit in the storm, a god, very like his powerful enemy, too, in the thunder, a nymph in the stream, a Pan in the wood; and every mysterious happening in nature he attributes to the caprice of these spirits, or, if he happens to be a monotheist, to the caprice of the one Great Spirit. And further, he begins to try to appease Nature, to try to get his gods, or his god, if he has but one, into a favorable mood instead of a hostile one toward himself. To do this he begins to sacrifice to Nature. And when his want is very urgent he sacrifices the best that he has—his daughter or his son. Human sacrifice apparently has been practiced by most, if not by all, primitive people. It certainly was practiced all over the Mediterranean area. You read it in the Homeric stories, where you find that Iphigenia, daughter of Agamemnon, had to give her life to propitiate the gods when the Achaeans pointed their ship toward Troy. You read it in the story of Tyre and Sidon where the Phoenicians offered up their children to Moloch. You find it in Palestine, where Abraham started to offer up his son Isaac.

Now look at the first forward step in the evolution of religion. Somebody arises somewhere, somehow, who begins to do a little reflecting on his own account. In the Bible story it was Abraham who began to

wonder whether Nature was after all just
a powerful, cruel, vengeful brute, like the
king of the adjacent tribe, who delighted
in, or was appeased by, human blood;
whether, in other words, the real God was
a being who could be propitiated by the
sacrifice on the part of a father of his only
son. And he answered no, and decided
then and there to break with the past.

The Bible story says, "God spoke to
Abraham." How did he speak? Through
some Arab sheik who just then passed that
way? Then it was the sheik rather than
Abraham to whom God spoke. Through
a voice that would have left a record on a
phonograph concealed in the bushes? Who
wants such a childish interpretation? Or
was it through the still small voice of reflection? But even so, where did that idea
come from that got into Abraham's mind?
I do not know. The most amazing thing
in all life, the greatest miracle there is,
is the fact that a mind has got here at all,
"created out of the dust of the earth."
This is the Bible phrase, and science today
can find no better way to describe it—a
mind that just *begins* to think for itself,

to relate phenomena and to that extent to understand them, to grasp a little of the mystery of existence, to make choices, to exercise intelligence. Where do our ideas come from? I do not know. All that we know is that somehow we are here, and, most wonderful of all, that we *know* that we are here, and that sometimes great new conceptions that lead us on to better things spring up in the minds of men.

"God spoke to Abraham." I do not know any better way in which the modern man can put it, and certainly primitive man with his animistic and anthropomorphic conceptions literally had no other way in which he could have stated it. God spoke to Lycurgus, too, when the Spartan lawgiver ordered human sacrifices stopped in Sparta, and at a time not many centuries after that at which Abraham had them stopped in Palestine.

Abraham and Lycurgus were very much alike, too, in that neither was able to do more than to take a little, faltering, first step in getting away from this manlike or anthropomorphic conception of the deity. Abraham thought that God might

not be pleased with the sacrifice of his son
Isaac, but that he would be pleased if
Isaac were replaced by a goat or a sheep,
and a whole religion grew up around the
notion that God, or the gods, could be propitiated with the sacrifice of animals. All
over the Mediterranean area we find this
practice continuing for hundreds of years.
Lycurgus did not go quite as far as Abraham, for when he got the definite conception that God was not pleased with human
sacrifice he replaced it by the flogging of
young men, still trying to propitiate God
by human suffering, though no longer by
the killing of the sons and daughters of
Sparta. With the abolition of human sacrifice, then, the first stage is passed in the
evolution of religion.

Before going on to the second stage a
definition is needed to avoid misunder-
standing. What is religion, as I am using
the word? Historically I think that religion has always dealt with two groups of
ideas, first, with one's conception of the
meaning of existence, of what is behind
these various phenomena of life, coördi-
nating them and giving unity and signifi-

cance to nature—in a single word, with his conception of God, and, second, with his conception of his own responsibility in this world, with his own place in the scheme of things. This last idea obviously grows out of the first, and is inevitably intertwined with it. These two ideas have always been associated in all religions, namely, ideas about the nature of God and definite notions about duty and responsibility. But now notice how these *conceptions* of *God* and of *duty* change as man learns more and more and gets farther and farther away from the earliest stages of his development.

We are now ready to turn to the second stage of the evolution of religion. Millions of people have lived since the preceding stage. They have all lived conventional lives; they have all done what their neighbors were doing. They have all brought the first-born of their sheep and cattle and goats to be sacrificed to their God or their gods. They have all paid "tithe of mint and anise and cummin." In a word, though they no longer believe in a God who demands human sacrifice, their conception

of the Deity is still extraordinarily manlike. Their God is a being who takes pleas-
ure in the smell of the sacrifices of beasts;
a being who can condemn whole families
and whole nations to destruction because
some member of the family or nation has
incurred his displeasure—this even in
Judea. In Greece the gods are still nothing
but overgrown, petulant children, with
magical powers but nothing else to dis-
tinguish them from humans.

And then another divine event occurs,
divine in just the same sense as the last.
A new idea comes into human thought
and life. It comes in a very limited way
through Mohammed, in a much larger
way through Buddha, in a great big swell-
ing tide through Jesus—a new conception
of God. Jesus struck the most mortal blow
that has ever been struck at all childish
literalisms, at all the ideas which underlie
modern so-called fundamentalism, when
he changed the literalistic interpretation
of the Jewish scriptures, the anthropomorphic conception of God prevalent up
to his time, and saw in God no longer
merely a powerful human being, but a

being whose qualities transcended all
human qualities; when he cried, "It hath
been written . . . *but* I say unto you";
when he saw a great benevolence behind
the universe; when he taught, "God is a
spirit"; when he said, "The kingdom of
heaven is within you"; when he, for the
first time in the history of the Jews, con-
ceived a God who was not interested in
Judah or Israel alone, but whose sym-
pathies, whose benevolence, stretched out
through all the world; when he also
changed man's conception of duty, for
this always must change with the change
in his conception of God; when he focused
attention upon the Golden Rule rather
than on sacrifices and burnt offerings;
when he directed man's thinking, as he
inevitably had to do with his conception
of God, upon the duty of benevolence, of
altruism among men, the duty of seeking
the good of the whole instead of being
governed by self-seeking and greed, such
as possessed the souls even of the gods of
the olden time. His gospel was simply the
gospel of a beneficent creator whose most
outstanding attribute was love, and that

conception of course made love, unselfishness, the first duty of man. And through all the next thousand years of horrible strife and disaster the life and the spirit and, to an extent, the conception of Jesus was kept before the whole western world as the basis of its religion.

I would not at all overlook the backward steps which religion took during this period, for let us frankly admit that it did take backward steps. It became deeply encrusted with superstition. Jesus himself tried his utmost to get his followers away from the idea that his authority rested upon any miraculous event, any caprice in nature. His kingdom, as he repeatedly asserted, was in the hearts of men, and he refused to let his disciples build it upon a sign. But his followers had not risen to his height. In that animistic age unusual events could not possibly be described or understood save in terms of "possession of demons" and the like. It is no wonder that the followers of Jesus during the next fifteen hundred years based their religion so largely as they did upon signs and wonders, and that the beautiful life and

teachings of Jesus had to shine through a
great mass of superstition which Jesus
himself, certainly to a much larger extent
than his successors, had tried to get away
from. Ideas that permeated all Mediterranean society could not possibly be elimi-
nated in a year or in a thousand years.

Then, about fifteen hundred years later,
another great, new step begins to be taken.
If one is to connect this step with any
one name it is with the name of Galileo
that we must associate the introduction
as a ruling principle in life of the scientific mode of thought. These new conceptions were not unknown to Aristotle and
the other most outstanding intellects of
Athens and of Alexandria, but it is only
from the time of Galileo that they begin to
modify in an enormous way the whole
world's conception of what this creation
of which we are a part is like,—in other
words, the world's conception of God and
the way in which he works.

What was Galileo's method? As indi-
cated in the preceding lecture, he lived in
an age when people altogether naturally
followed the teachings of Aristotle with

respect to the relations of force and motion, but Galileo, just like Abraham and just like Lycurgus, began to question the correctness of the conventional belief, and instead of being content with inherited hearsay knowledge he said to himself, "I will try by careful experiments to see whether it is correct or incorrect." That is how he came to make the famous experiment at the Leaning Tower of Pisa as a result of which the formula which had been accepted for two thousand years by millions of people could be accepted no longer.

But Galileo went further still. He was not interested in merely destructive criticism, in the philosophy of knock, for this is one of the easiest philosophies in the world. It requires not one whit more intelligence than does conventional morality. The unthinking follow both of them in crowds. But Galileo's was of a higher order of intelligence. He was a constructive thinker. By years of patient effort and careful experiments he now sought to replace the old erroneous conception by a correct one, and as the result of a lifetime of effort he introduced into the world, as has been already detailed, the idea that is stated in Newton's second law of motion that force or effort is proportional not to motion imparted, but to rate of change of motion. In the usual formulation, force is equal to mass times acceleration.

I said in my last lecture that no idea that has ever come into human thought has exercised so profound an influence upon the development and destinies of the race as has this idea of which I am speaking. I said this, not primarily because our whole modern material civilization depends upon it, but rather because the scientific method by which Galileo got at his new idea began at about this time to change in a large way the whole mode of thought of the human race, to change the philosophic and the religious conceptions of mankind, because the foundations were here laid for a new advance in man's conception of God, for a sublimer view of the world and of man's place and destiny in it.

Jesus had gone a long way toward destroying or refining man's primitive, childish conception of a capricious, anthropomorphic God. The method of Galileo, worked out through the following centuries, took a further step in the same direction. It began to show us a universe of orderliness and of the beauty that goes with order, a universe that knows no caprice, a universe that behaves in a knowable and predictable way, a universe that can be counted upon; in a word, a God who works through law. Yes, more even than that, a universe that is not only willing to let us know her, but that has untold forces and powers which can be counted upon to work for the benefit and enrichment of human life as soon as we can learn to understand them and to work in harmony with them. It was useless to try to do this so long as God was a capricious being.

Here was another divine event, the third stage in the evolution of man's conception of God and, as an inevitable consequence, of his conception of duty. The monasteries of the middle ages testify to the old conception of God and of duty; the insistent, effective activity of a Maxwell, a Pasteur, or a Kelvin to find out the laws of nature and to turn them to the amelioration and the enrichment of human life, to the new. The new God was the God of law and order, the new duty to know that order, and to get into harmony with it.

These new ideas of course were not attained all at once. They grew and spread slowly from about 1500 up to the present time, and culminated in what I shall call the fourth stage in the evolution of religion, the stage in which we now are—a stage that is ushered in through the growth of another sublime idea or through a new revelation of God to man, in the idea that has come into human thinking out of the utilization of Galileo's method in the study of geology, of biology, of physics, of paleontology, of history, an idea in the development of which Darwin has been one of many outstanding figures.

Through the careful study of the way the rocks lie on our hillsides we have found evidence for the growth of this earth through a billion years at the least. Through the study of radioactivity and other physical processes we have found definite evidence that the world is evolving and changing all the time, even in its chemical elements. By a minute study of the comparative anatomies of all kinds of animals and by reading the history of life through fossils we have found evidence of progression, evidence of a continuous movement from the lower up to the higher forms, and through the study of history and the observation of what is going on under our eyes at the present time a new conception, *a conception of progress, has entered the thought of the world,* a progress in which we play an important part, a progress the key to which is to a considerable extent, at least, in our own hands. The picture which the development of science and the scientific method has brought into the world of a continual increase in control over environment is the dominant note in the fourth stage in the evolution of religion. No conception of God which has ever come into human thinking has been half so productive of effort on the part of man to change bad conditions as has this new modern conception of progress, this conception that man himself plays a part in the scheme of evolution, this conception which has arisen because of work like that of Galileo, like that of Pasteur, and especially like that of Franklin and Faraday, that it is possible in increasing measure for us to know and to control nature; this conception, inevitably introduced into human thinking by the stupendous strides which have been made in the past century, that there are perhaps limitless possibilities ahead through the use of the scientific method for the enrichment of life and the development of the race.

In this sense the idea that nature is at bottom benevolent has now become well-nigh universal. It is a contribution of science to religion, and a powerful extension or modification of the idea that Jesus had seen so clearly and preached so persistently. He had *felt* that benevolence and then preached it as a duty among men. Modern science has brought forward *evidence* for its belief. True, it has changed somewhat the conception and the emphasis, as was to have been expected, for it is this constant change in conception with the advance of thought and of knowledge that we are here attempting to follow; but the practical preaching of modern science —and it is the most insistent and effective preacher in the world today—is extraordinarily like the preaching of Jesus. Its keynote is service, the subordination of the individual to the good of the whole. Jesus preached it as a duty— for the sake of world-salvation. Science preaches it as a duty—for the sake of world-progress. Jesus also preached the joy and the satisfaction of service. "He that findeth his life shall lose it, and he that loseth his life . . . shall find it." When the modern scientist says he does it "for the fun of it" or "for the satisfaction he gets out of it," he is only translating the words of Jesus into the modern vernacular. It would be hard to find a closer parallel.

Concerning what ultimately becomes of *the individual* in the process, science has added nothing and it has subtracted nothing. So far as science is concerned religion can treat that problem precisely as it has in the past, or it can treat it in some entirely new way if it wishes. For that problem is entirely outside the field of science

now, though it need not necessarily always
remain so. Science has undoubtedly been
responsible for a certain change in religious thinking as to the relative values of
individual and race salvation. For obviously by definitely introducing the most
stimulating and inspiring motive for altruistic effort which has ever been introduced, namely, the motive arising from
the conviction that we ourselves may be
vital agents in the march of things, science
has provided a reason for altruistic effort
which is quite independent of the ultimate
destination of the individual and is also
much more alluring to some sorts of minds
than that of singing hosannas forever
around the throne. To that extent science
is undoubtedly influencing and changing
religion quite profoundly now. The emphasis upon making this world better is
certainly the dominant and characteristic element in the religion of today. Nor
is it confined to the formal religious organizations, though it probably gains its
chief impulse from them. For this new
idea of progress, and of our part in it, and
our responsibility for it, is now practically

universal. Call it an illusion if you wish, but you at least cannot deny the existence of the idea, and it is *ideas* that count in this world, for in them is, of course, the motivation of all conduct. For my own part, I am going to call the introduction of this idea as divine an event as has ever taken place. It is due directly to science, and it marks the latest stage in the evolution of religion, *i.e.*, the latest stage in the evolution of man's conceptions about the ultimate nature of his world and his relations to that world—his conceptions about God and about duty.

In the midst of these changing conceptions there are of course crowds that hang behind, that cannot break away at all from the traditions and the life of the past, and there are of course other crowds that want to break completely with it, that call it all a "pack of lies," that have not enough discernment to see the truth of the past unless it wears the precise garb and hue of the present. *Neither of these two crowds has any conception of what the evolutionary process means.*

It is not a question of whether one is

religious or irreligious, so much as whether one is scientific or unscientific, rational or irrational. The world is of course "incurably religious." Why? Because everyone who reflects at all *must have* conceptions about the world which go beyond the field of science, that is, beyond the present range of intellectual knowledge. As soon as we get beyond that range we are in the field that belongs to religion, and no one knows better than the man who works in science how soon we get beyond the boundaries of the known. These boundaries are continually changing, and so the conceptions that must start from them, and have their footings in them are likewise of necessity changing. That is, religion is changing *now,* because of the interplay of science upon it, precisely as it has been changing in the past, and especially during the past century.

As I see it, there are but two points of view to be taken with respect to this whole question of religion. The one is the point of view of the dogmatist; the other the point of view of the open-minded seeker after truth. Dogmatism means assertive-

ness without knowledge. The attitude of
the dogmatist is the attitude of the closed
mind. There are two sorts of dogmatists in
the field of religion. One calls himself a
fundamentalist; the other calls himself an
atheist. They seem to me to represent
much the same type of thinking. Each asserts a definite knowledge of the ultimate
which he does not possess. Each has closed
his mind to any future truth. Each has a
religion that is fixed. Each is, I think, irrational and unscientific. The fundamentalist is so because in his assertiveness about
the ultimate and the unknown, he trenches
on the known, and asserts as true that
which we now have every reason to believe
is false, such as the six-day creation of the
earth or the rotation of the sun about it.
The atheist, on the other hand, is irrational and unscientific because he asserts
that there is nothing behind or inherent in
all the phenomena of nature except blind
force, and that in the face of the fact that
he sees evidence of what he is wont himself
to call intelligence in the workings of his
own mind, and in the myriads of other
minds which are a part of nature. The only

way I see to relieve him of this charge is
to assume that he uses words such as
"atheist" and "blind force" in a sense entirely different from that in which everybody understands them, and this itself is
unscientific. The God of science is the
Spirit of rational order, and of orderly
development. Atheism as I understand it
is the denial of the existence of this spirit.
Nothing could therefore be more antago-
nistic to the whole spirit of science. Even
Voltaire condemned it as unintelligent
when he wrote: "If God did not exist it
would be necessary to invent him." If I
myself were confronted with a choice between these two types of dogmatic religion, fundamentalism, and atheism, and
could not find a way to take to the woods,
I should choose fundamentalism as the less
irrational of the two and the more desir-
able, for atheism is essentially the philosophy of pessimism, denying, as it does,
that there is any purpose or trend in
nature, or any reason for our trying to
fit into and advance a scheme of development, and any such denial is a direct con-

tradiction of the evolutionary findings of all modern science.

But fortunately I am not obliged either to make the foregoing choice or to take to the woods; for there is another kind of religion—a religion which keeps its mind continually open to new truth, which realizes that religion itself has continuously undergone an evolution, that as our religious conceptions have changed in the past so they may be expected to change in the future, that eternal truth has been discovered in the past, that it is being discovered now, and will continue to be discovered. That kind of religion adapts itself to a growing, developing world. It is useful in such a world while both kinds of dogmatic religion seem to me to be useless. If the present organizations of religion in the churches can adapt themselves to, and keep pace with, our continually increasing knowledge, they will continue to be one of the most potent factors in our progress. If they cannot do so they will be swept aside into the backwash of the current of progress and some other organization will be formed to do their work, *for religion will be with us so long as man hopes and aspires and reflects upon the meaning of existence and the responsibilities that it entails.*

Thus far I have been dealing with the changes in religious conceptions that have accompanied, and been occasioned by, the growth of the race in knowledge, and we have found these changes very like those which accompany our own thinking about Santa Claus as we pass from childhood to maturity. At the age of four Santa Claus, with his whiskers and his pack and his fifty-inch waistband, was on Christmas eve the most real being in the world. By the age of seven or eight we had measured up the chimney and found it woefully inadequate for the fifty-inch waistband, and Santa Claus became a myth, unless, perchance, we happened to be born of wise parents. By the age of twelve or fifteen, if we were fortunate enough to have younger brothers and sisters, he had begun to come back, and at thirty, when we were hanging the stockings of our own little ones, Santa Claus—the spirit of Christmas—was more real than he had ever been in childhood, and the eight-inch chimney and the fifty-inch waistband no longer mattered.

We have just learned, to our amazement, through the fundamentalist movement, that a very considerable portion of America is still in the four-year-old stage of its religious development. We are not so much surprised to know that many more of us are in the seven-year stage, for we have had a group of blatant writers to remind us of that right along. Indeed, across the water this stage of development is wont to be regarded as most typical of America. G. Lowes Dickinson makes Ellis say in *A Modern Symposium,* "Thanks to Europe, America has never been powerless in the face of Nature, therefore has never felt fear, therefore never known reverence, therefore never experienced religion." A recent dialogue clipped from the London *Spectator,* issue of April 17, 1926, is accurately descriptive of the bumptious self-confidence, the undiscerning irreverence of this second stage.

ADAM AND EVE

EVE:

What should we do, love, if the sun should fail,
(There have been times when he grew wan and
  pale)
If he his daily task should not complete,
Nor give his kindly boon of light and heat?
Some day he may be weary and foredone—
What shall we do if we outlive the sun?
And those frail, pretty stars, and that weak
  moon,
Surely their strength will be exhausted soon;
How we shall grieve when they have spent their
  light,
How we shall miss them from the sky at night!

ADAM:

Vex not your thoughts about yon flaming ball;
I'll find another should it fail or fall;
Borrow the eagle's wings that I may fly
And set it on its path across the sky;
Sojourn a little space in that high air
And put the stars to rights while I am there.

EVE:

How brave you are, my Adam—brave and wise,
More marvellous than the whole of Paradise.
Yet now I see my thoughts were foolish ones.
He who made you can make a thousand suns,
And He who rules the even and the morn
Can scatter stars as I these grains of corn.

ADAM (*rather annoyed*):

No, I shall see to it myself.

Rose Fyleman.

This sort of presumptuous, strident, blatant, undiscerning irreverence has had its most conspicuous representatives in America in recent years, not, however, among scientists, though it is sometimes thought to be the characteristic attitude of modern science, and possibly with a certain element of justice. Physics, however, has recently learned its lesson, and it has at the present moment something to teach to both philosophy and religion, namely, the lesson of not taking itself too seriously, not imagining that the human mind yet understands, or has made more than the barest beginning toward understanding the universe. Today physics is much more open-minded, much less dogmatic, much less disposed to make all-inclusive generalizations, and to imagine that it is dealing with ultimate verities, than it was twenty-five years ago. This generalizing farther than the observed facts warrant, this tendency to assume that our finite minds have at any time attained to a complete understanding even of the basis of the physical universe, this sort of blunder has been made over and over and over again in all periods of the world's history and in all domains of thought. It has been the chief sin of philosophy, the gravest error of religion, and the worst stupidity of science—this assumption of unpossessed knowledge, this dogmatic assertiveness, sometimes positive, sometimes negative, about matters concerning which we have no knowledge. If as we pass from the seven-year-old to the thirty-year-old stage of our racial development our conceptions of God become less childishly simple, more vague and indefinite, it is because we begin to realize that our finite minds have only just begun to touch the borders of the ocean of knowledge and understanding. "Can man with searching find out God?" If there is anything that is calculated to impart an attitude of humility, to keep one receptive of new truth and conscious of the limitations of our understanding, it is a bit of familiarity with the growth of modern physics. The prophet Micah said, twenty-five hundred years ago, "What doth the Lord require of thee but to do justice, to love mercy, and to walk humbly with thy God?" Modern science, of the real sort, is slowly learning to walk humbly with its God, and in learning that lesson it is contributing something to religion.